An Industry Now Lost

The Pride, Passion and Pain of Mining

ARTHUR COLE

WORDCATCHER publishing

An Industry Now Lost
The Pride, Passion and Pain of Mining
Wordcatcher Modern Poetry

© 2018 Arthur Cole

Cover photograph of the 9-foot high sculpture *The Keeper of the Collieries*, Maesteg, sculpted by Chris Wood © 2018 Arthur Cole
Background cover image supplied by Adobe Stock
Cover design © 2018 David Norrington

The Author asserts the moral right to be identified as the author of this work. All rights reserved. This book is protected under the copyright laws of the United Kingdom. Any reproduction or other unauthorised use of the material or artwork herein is prohibited without the express written permission of the Publisher.

No part of this book may be reproduced, stored in a retrieval system, or transmitted in any form or by any means, electronic, electrostatic, magnetic tape, mechanical, photocopying, recording or otherwise, without the written permission of the Publisher.

British Library Cataloguing in Publication Data.
A catalogue record for this book is available from the British Library.

Published in the United Kingdom by Wordcatcher Publishing Group Ltd
www.wordcatcher.com
Tel: 02921 888321
Facebook.com/WordcatcherPublishing

First Edition: 2018

Print edition ISBN: 9781789420210
Ebook edition ISBN: 9781789420203

Category: Poetry

Contents

Never to be Forgotten 1

Aberfan 2

Parc Slip Remembered 4

When Coal Was King 6

Gresford Colliery Disaster 8

Under the Ocean 10

A Widow's Tale 12

Death Under the Arctic 13

A Tribute to Tower... Glofa Twr 14

A Legacy of Coal Mining 16

A Wife's Prayer 17

No Pit For You Boy 18

Trappers, Hurriers and Thrusters 19

Garden Pit Remembered 20

The Ghost with the Hammer in His Hand 21

The Thirty-three 22

Ferndale... Black Gold 23

Pit Ponies 24

A Coal Miner's Wife 25

Dust to Dust 26

To Sir With Love... Mr David Beynon 28

A Canary's Tale 29

The Blue Scars... 30

Brave Men of Gleision... 31

Cathedral.. 32

The Runaway 33

Kellingley – Last of an Era... 34

Now But a Memory – Caerau 36

Senghenydd Remembered.. 37

The Sinking Pit 38

Ynyscedwyn-Hendreladus Colliery Disaster, 24.08.1892 40

The Feeder and Suckers 41

Put the Coal Away... 42

Traits of a Miner 43

The Knockers... BWCA 44

Disaster in the Trefor 45

Weekly Card School... Albert Street, Caerau..................... 46

Lord Save Me Now... Arglwydd achub fi nawr.................. 48

Tip Number Seven 50

Tragedy in the 'Silkstone Seam' 51

Keeper of the Collieries 52

I Wander the Face.. 53

Always in our Prayers 54

Was it Their Destiny? 56

Father and Son in Spirit 58

Pit of Nations – Gedling Colliery.. 60

Knockshinnock Castle Colliery Mining Disaster 62

The Eden Colliery – Place of Pleasure 64

A Pony's Tale.. 66

Hope… Gleision 67

Never to be Forgotten

A generation lost, a valley's heart broken,
fifty years on, many memories unspoken.
A whole nation mourned, that autumnal morn,
'Our Angels' resting place, beautifully adorned.

We must never forget that October dawn,
the world stood in silence, fifty years on.
Will time heal, only loved ones can say,
to turn it back, they pray each single day.

'Our Angels' together in the Lord's hands,
at peace forever, guided by his holy command.
Let's not forget, the protectors that day,
they gave their lives, on that morning so grey.

Survivors now living, God bless and protect,
our childhoods stolen, such a profound affect.
We will never forget 'Our Angels' and you,
a generation lost, rest in peace where they grew.

This poem was written to commemorate the fiftieth anniversary of the Aberfan disaster. R.I.P. our angels. Peace and love to all those that survived.

Aberfan

Slag and slurry, the devil incarnate,
robbed a village of lives and didn't abate.
On that early morn, the sun fell from the sky,
a giant black shadow drowned their small cries.

Teachers' and children fighting for air,
the slag and the slurry laid the school bare.
Completely covered by the river of black,
many souls lost, if we could only go back.

The village would come, with all haste and speed,
digging with hands, miners taking the lead.
Mothers wailing, where is my child?
The black stuff still sliding, a torrent so vile.

The number it took was one forty-four,
if only God had allowed a few hours more.
Half term it would be later that day.
souls that we lost, would have been on their way.

The survivors they numbered one forty-five,
sad and heartbroken, however, alive.
God bless you all, please rest in peace,
never forgotten, our love will not cease.

At 9.15am on Friday 21st October 1966, a colliery spoil tip collapsed into many homes and the Pantglas Junior School, killing 116 children and 28 adults. This poem is dedicated to them. May they all Rest in Peace.

Parc Slip Remembered

One hundred and twelve, at peace in heaven,
in their day the heartbeat of Cefn.
Parc Slip the mine where they all worked,
way down below, their destiny lurked.

Men and boys taken that day,
firedamp and flame, the usual way.
Generations lost, without any say,
deadly fire and gas, took all air away.

Two miners were rescued, after the blast,
no hope for the rest, the future now past.
Hampered by falls, the rescue continued,
the following day, hope came to the village.

Voices were heard, deep in the mine,
a miracle, rescued, at least thirty nine.
Some solace for those that were weeping,
heart break for those, loved ones sleeping.

Still to this day, Cefn mourns the begotten,
those lost souls, never forgotten.
Descendents reflect, heartache and sorrow,
their relatives felt on that sad 'morrow.

Parc Slip is now a nature reserve,
a memorial, those heroes deserved.
Birdsong is heard each single day,
lost souls at last finding their way.

At 8.20am on 26th August 1892, an explosion occurred at the Parc Slip Colliery, Aberkenfig, near Tondu. One hundred and twelve men and young boys lost their lives, may they all Rest in Peace.

When Coal Was King

Brave colliers, his loyal subjects,
green valleys were his throne.
Miles underground his heart beat,
those days, now so forlorn,
When coal was king.

Villages throbbed with life back then,
the club, the pub, the miner's hall,
Chapels filled with riotous children,
front steps all scrubbed with pride,
When coal was King.

Lumps and small dropped in the lanes,
to keep the families warm,
A warm tin bath by the fire,
backs scrubbed by loving wives.
When coal was King.

Families struggled, back in the day,
brave men, maimed and killed.
Widows, children, left without hope,
their future a royal legacy,
When coal was King.

From boys to men, with box and jack,
two miles down, their second home.
Young men old before their time,
black dust their holy maker,
When coal was King.

Today the valleys, tired and bleak,
the King has lost the throne,
we can only dream of times gone by,
fading memories of the past.
When coal was King.

Gresford Colliery Disaster

It was an early September morning,
brave colliers were earning a crust.
'twas then that the Dennis erupted,
methane and fire took their souls.

Two sixty-three, dispatched to heaven,
families never to see them again.
Heartbreak and sorrow would follow,
owners lying, compounding the pain.

This disaster was a sign of the times,
owners were always driven by profits.
A collier's safety, nowhere to be found,
ventilation, fresh air they were secondary.

To rescue these lost souls proved futile,
three brave men dispatched to their maker.
Two sixty-six, was now the total,
the Dennis a tomb now forever.

This disaster's cause was never proven,
owners would get off scot free.
Yes, a life underground it meant nothing,
it was all about money and greed.

The Dennis now a tomb underground,
no gravestones for families to mourn.
All that is left are the memories,
of the sacrifice they made for black gold.

In memory of the 266 lost souls who perished in the explosion on 22nd September 1934 at the Gresford Colliery, near Wrexham, North Wales. Eleven bodies were recovered, the rest were entombed.

Under the Ocean

His gramps was a miner for forty-five years,
from a boy to a man, had shed many tears.
Saw many 'butties' injured and broken,
'no pit for you boy', was so aptly spoken.

Garwyn pestered his gramps, please take us down,
fourteen we were, likely lads about town.
An adventure we thought, a walk in the park,
our outlook soon changed, once down in the dark.

Garwyn's gramps agreed, permission granted,
hard hat and lamp, got what we wanted.
Down in the cage, pit bottom we land,
walls all whitewashed, at grampa's command.

The ponies were there, stabled and fed,
miners all smiling and nodding their heads.
Through the large doors and into the face,
steel rings and props, a warm humid place.

Coal dust in the air, hard men cutting coal,
stripped to the half, baring bodies and soul.
Conveyors running, drams being filled,
like black ants at work, a sight to behold.

Back up in the cage to the fresh air and light,
our faces all black, hearts filled with delight.
Dust I could taste at the back of my throat,
a day to remember, one worthy of note.

Now as a man, I reflect now and then,
under the Ocean, with all those brave men.
Mines in our valleys, are sadly now gone,
a memory forever, in my mind will live on.

> *Written for a good friend of mine, who wanted to recollect a visit he had made to the Ocean Colliery, Blaengarw with his best friend when they were young.*

A Widow's Tale

We married young, had the world at our feet,
a family we planned, to make life complete.
Heartbreak and sorrow now with me for life,
a widow I am, no longer a wife.

That fateful morning, songbirds on high,
not a care in the world, a cloudless blue sky.
His box I prepared with a jack full of still,
a cwtch and a kiss, a wish of goodwill.

A few hours later the hooter did blast,
I knew in my heart, that life had been lost.
Women all rushing, children in tow,
panic and prayer for their loved ones below.

Atop of the mine, men with faces like stone,
down below rescuers, working like drones.
A fall was the cause, three miners entombed,
the love of my life, would be later exhumed.

The hours they passed, brought up one by one,
bodies bleeding and broken, a bright morning sun.
My husband and lover, a blanket did cover,
no cwtches and kisses, our love affair over.

That morning I cried, my heart broken in pieces,
from a wife to a widow, that grief never ceases.
He died in the bowels below, with butties he loved,
no solace for me, left alone up above.

Death Under the Arctic

The Arctic circle, so desolate, so cold,
under its ground, thirty lost souls.
Vorkuta mourns her brave men below,
entombed forever, with no place to go.

Explosions abounded, miners in prayer,
methane the killer, polluting the air.
Twenty-six trapped, four killed outright,
daylight no more, for life they did fight.

Another explosion, the fires, they raged,
six more souls lost, no rescue to stage.
In all thirty-six, at Severnaya were lost,
coal they were mining, at a terrible cost.

Nitrogen gas was used to put out the flames,
ironic this fact, as methane was to blame.
Families will mourn the rest of their lives,
precious, fond memories, for lovers and wives.

In that hell hole below, a breed set apart,
bonded together, in body and heart.
Russia will mourn, those brave men below,
high in the sky the Northern Lights glow.

Between the 25th and 29th February 2016, thirty-six were killed in the Severnaya coal mine, Vorkuta, Komi Republic, as a result of explosions. The cause is believed to be leaking methane gas.

A Tribute to Tower... Glofa Twr

Crawshay's folly, gave Tower its name,
a world famous pit, it later became.
Coal was abundant, the valley alive,
plenty of work, families could thrive.

Then Maggie came wielding the axe,
deep mines would close and that was a fact.
Tower was listed pre eighty-four,
new roads driven, gave her ten years more.

The death knell came in ninety-four,
Tower succumbed to Maggie's law.
The valley weeping and so forlorn,
brave men so angry, venting their scorn.

What now the future, a community spent,
no work for the men, so much discontent.
A buy-out the plan, Tower is viable,
Tyrone led the way, a man so reliable.

Like a phoenix she rose, producing again,
the valley alive, after oh so much pain,
Men again mined, a mistress they loved,
always protected by the good Lord above.

For fourteen years, the Tower stood proud,
then the day came, another black cloud.
The closure now final, but not by the Tories,
the valley all marched and regaled in their glory.

The story of Tower will live on forever,
a legacy of hope, faith and endeavour.
This special breed, remembered with pride,
families of heroes, who fought side by side.

A Legacy of Coal Mining

Deep in the bowels, on stomachs and knees,
a lifetime of dust and that deadly disease.
Black gold they hewed, from morning till night,
a band of brothers, they fought the good fight.

Young men growing old, in double quick time,
older men coughing, cut off in their prime.
A strong back you needed back in the day,
now and again they would silently pray.

Working in water all through a shift,
no good for the joints, arthritis the gift.
The constant noise, a deafening din,
blue scars, like medals they wore on their skin.

They say hard work, it never hurt,
I say, ask those men who have dug in the dirt.
Old they became, before their time,
a mistress they served, a love still sublime.

A Wife's Prayer

Lord guide him, keep him safe,
as he descends the bowels of the beast.
A man of courage, a man of strength,
energy and spirit, rarely matched.
Lord keep him safe.

A brother bonded by 'black gold',
kneeling, crawling, groping in the dark.
Like a blind beggar, seeking alms,
his life below, in suspension.
Lord keep him safe.

The monster, lying dormant, 'til disturbed,
spewing dust, to all corners of the face.
Methane gas, monoxide, silent killers,
lurking in every seam and crevice.
Lord keep him safe.

Another shift, no loss of life,
you brought him home, to his safe haven,
I watch him sleep undisturbed,
I thank you from the bottom of my heart.
Lord you kept him safe.

No Pit For You Boy

A miner he'd been from a very young age,
a kind loving man, with the odd bout of rage.
'No pit for you boy' he'd always say,
schoolings the way to earn tidy pay.

Being so young, I took no heed,
never missed school, so could write and read.
I played the fool and was full of wit,
at sixteen I left, followed dad down the pit.

He wasn't best pleased, and that is a fact,
flew into a rage, didn't hold back.
Once he'd cooled down, came back to earth,
he gave me his blessing, I could see he was hurt.

For ten years and more, worked side by side,
went from a boy to a man, no easy ride.
Got married, had sons, life wasn't easy,
but the pit was our life, my legacy.

My dad's health declined over the years,
dust in his lungs, eyes welled with tears.
Fighting for breath, in agony and pain,
his life at an end, but not in vain.

Dad finally passed, peaceful one night,
I remember those words, yes he was right.
'No pit for my boys' I whispered to him,
it's schooling for them, not dust and black phlegm.

Trappers, Hurriers and Thrusters

A trapper I was, working twelve hours a day,
down a pitch black hole, is where I would lay,
Trap doors I opened, to let the air flow,
while coal tubs travelled back to and fro.

Never alone, down that deep dark pit,
my family all there, doing their bit.
Hurriers and thrusters, my brothers both,
tubs pushed and pulled, for all their worth.

My mother, the same, she played her part,
tried to protect us, with all of her heart.
A thruster she was, strong as an oxen,
but times were hard, there was no other option.

Fathers and youths worked down in the deep,
with pick axes in hand, cut coal for our keep.
Holding only a candle to show them the way,
after twelve hours of pain, a pittance for pay.

Mining was hard in those Victorian times,
there were many lives lost, cut off in their prime.
New legislation, just had to be passed,
so families could survive, blood lines would last.

Women and children aged under twelve,
no longer black holes, they'd have to delve.
Times have moved on, I'd like to think,
trappers, hurriers, thrusters now all extinct.

Garden Pit Remembered

The Cleddau flowed with grace and ease,
so calm and tranquil, birds on the breeze.
Below they crawled not knowing their fate,
miners of Garden, it would decimate.

Forty brave miners, below in the gloom,
fate later sealed, a water filled tomb.
It started with seepage, early that morn,
no heed was paid, souls later forlorn.

The Cleddau did rage, through tidal flow,
pit walls they burst, a deluge below.
For safety they ran, the water gave chase,
engulfing the souls and filling the face.

Who was to blame for that fateful day,
owners renaged, because that was the way.
They treated miners with contempt and disdain,
money and profit, was the name of the game.

The Cleddau still flowing down to the sea,
was she to blame for this tragedy.
She's guided by nature, that we all know,
It's the owners I blame, as history shows.

The Ghost with the Hammer in His Hand

A Quakers Yard boy, born and raised,
heart of a lion, the world he'd amaze.
The son of a miner, hard as teak,
small of stature, a nature freak.

At the age of twelve, the mine did call,
together with men, the gullies, he'd crawl.
After four years, the booths were his calling,
bare knuckle fighting, big men falling.

He then turned pro and fought for a living,
big or small, he was unforgiving.
The mighty atom, the name he was dubbed,
world flyweight champion is what he'd become

For sixteen years, he fought with no fear,
still to this day, he has no boxing peer.
Many have followed, many have tried,
this legend lives on, our nation's pride.

For years, enduring all punches hurled,
a miner's son, he conquered the world,
Not a man to taunt or rile,
that was the legend Jimmy Wilde.

The Thirty-three

The San Jose mine, a disaster in waiting,
families alone, left, hoping and praying.
Thirty-three miners, entombed down below,
the treasure they sought, copper and gold.

The mine had a history of fires and falls,
the owners however, paid no heed at all.
Half a mile down, five miles from the entrance,
a cave-in occurred, brought a death sentence.

A rescue was mounted, no hope it was thought,
but the people of Chile, rallied and fought.
Boreholes were drilled, rescuers praying,
no sound, just silence, dead they were saying.

After seventeen days, God answered the prayers,
a note on a bit, we're alive don't despair.
Families cried, rescuers cheered,
food and water, is what they now need.

For fifty-two days, the drilling resounding,
with winches in place, rescuers abounding.
The time had came to bring them home,
sixty-nine days, they'd been all alone.

An emergency shelter had been their saviour,
for sixty-nine days, bravery, endeavour.
Like miners all over, a breed on their own,
courage and valour, this time not alone.

Ferndale... Black Gold

Nine mines there were, made Ferndale proud,
steam coal 'BLACK GOLD' up from the ground.
Seven seams, producing quality coal.
those were the days, no men on the dole.

In 1817, an explosive sound,
burying men, deep underground.
Choking and burning, screams would abound,
it took nearly a month, for them all to be found.

The cause it was found, safety locks tampered,
chances taken by miners, whilst being hampered.
A jury concurred, it was management's fault,
knew of the practice, never called a halt.

Two years later another explosion,
fifty-three souls, trapped, no ventilation.
Screaming and choking, in the bowels below,
what they went through, nobody knows.

The inquest verdict, bosses to blame,
no ventilation, no air in the seam.
If they had listened in '67,
Fifty-three souls, no place in heaven.

One other death occurred at Ferndale pit,
his name Thomas Charles he fought at Rorke's Drift,
He fought off the Zulu's with his rifle in hand,
later killed by a journey, in his promised land.

Pit Ponies

I was four when I started, my life underground,
stabled below, coal dust would abound.
Miners my friends, they treated me well,
for one, oh so young, the face was like hell.

The dust and the gas, the air putrified,
the miners would crawl on their bellies and sides.
After pulling the journeys for eight hours a day,
I lay in my stable, on soft and warm hay.

Fifty weeks of the year we'd work together,
think what I'd give for fresh air, fine weather.
Then it would come, two weeks on top,
roaming the fields, a nice gentle trot.

The air I took in, so fresh and clean,
the weeks would fly by, then back to the seam.
Ten years I would work, with the brave men below,
but my time it did come, up top I would go.

Up in the cage, to the top of the pit,
they patted my head, you deserve it.
Checked by the vet, then down to the field,
where for two weeks a year, always spring heeled.

A pit pony's life was hard and so tough,
I made many friends, took the smooth with the rough.
Life in the field, is the way it should be,
for ponies who started out young, just like me.

A Coal Miner's Wife

Strong of mind, tender of heart,
throughout their lives, a breed apart.
Stood by their men through thick and thin,
a duty performed, no arguing.

My mum, one of this rare loving breed,
looked after my dad, tended his needs,
Wages were low, housing was poor,
my mum was the glue, she kept us secure.

Brought us all up with values and manners,
looking back now, how did she manage.
How did she cope with all the despair,
strength, love and loyalty, qualities rare.

At the start of his shift, my dad's box she'd fill,
bread, jam and cheese, a jack full of still.
A kiss on the cheek, keep safe my love,
eyes closed and praying to God up above.

The wife of a miner, a breed apart,
so full of passion, a strong beating heart.
Many hearts broken over the years
widows in black, holding back tears.

Tough women they were, with a soft seam inside,
husbands and children, their love and pride.
Defying the odds, they got on with life,
There's no job compares, to a coal miner's wife.

Dust to Dust

Where did all those good days go,
flown by like birds on wing.
Memories, are all I cling to,
coughing and gasping for breath.

Once strong, so fit and robust,
a shadow, I have now become,
A few steps are all I can take,
the mask a lifeline for years.

I breathe like an old puffing billy,
the tank it's there on stand by,
Yes, all I have left are memories,
the seams of coal are long gone.

Forty I was when I finished,
my future cut short by dust.
Lungs filled with rubbish down under,
I haven't got long left to go.

My time on this earth I enjoyed,
lucky, I had butties for life.
I spit phlegm at the drop of a hat,
mask and tank, I need them right now.

Memories make me so happy,
a laugh just makes me cough.
I sit here and suffer in silence,
at my right hand is my wife

She worried for years about fires and falls,
our home a safe haven for both.
Twenty-four years of breathing in dust,
I would do it all over again.

When I go, I want you all to be happy,
cremation, that is my wish,
Ironic, I will be turning to dust,
the bastard that put paid to me.

To Sir With Love... Mr David Beynon

His calling to teach, nurture, protect,
loved by his pupils, so much respect.
That autumnal morn, all came to an end,
a mountain of slurry, death would descend.

Children so happy, innocent, no cares,
chatting like jackdaws, all unaware.
A thunderous roar, silenced and scared,
to Sir they ran, with no time to prepare.

A deluge swept down, unforgiving, so swift,
entombing the young, bringing death as its gift. To
Sir they ran, all fearing their lives,
he tried to protect, but with them he died.

Five children were found clutched to his chest.
a blackboard their headstone, their final rest. His
duty performed without favour or fear,
he tried to protect them, be their saviour.

David Beynon his name, that day a hero,
gave solace to five, knowing there was no tomorrow.
Children he nurtured, his love never ceased,
in heaven together, all now at peace.

Dedicated to Mr David Beynon, the deputy headmaster at Pant Glas Junior School, Aberfan. Mr Beynon was found dead, clutching and protecting five of his young pupils against the blackboard in their classroom. May they all Rest in Peace.

A Canary's Tale

Life could be short, back in the day,
hovering and singing, showing brave men the way.
Monoxide and methane, the killers below,
as long as we sang, onward they'd go.

Now and again, we'd breath in the poison,
our singing would stop, killer gas was the reason.
The silence would warn, methane, in the air,
run for your lives, a miner's nightmare.

Nineteen-eleven, our first use they found,
saving the lives, brave men underground.
Treated like pets, down there at the face,
sadly, however, bird and miner deceased.

In eighty-six, no more time at the face,
modern technology, us birds would replace.
The miners we saved, cannot be numbered,
our legacy, I hope, always remembered.

The Blue Scars

Those marks on your hands, 'What are they dad?'
'Don't worry, they're nothing' he said.
Like badges of honour, he wore them with pride,
I just stared, I was so mystified.

Whilst washing his back, I saw many more,
so deep all over his torso.
They bled for days, he never complained,
I thought my dad, was an absolute hero.

As I got older, spent more time with men,
those badges they all seemed to carry.
A bond for life I later realised,
their stories and lives in reprise.

The coal and the dust they made the scars,
a film they would make, some years later.
You all know its name, 'BLUE SCAR' it became.
even today, they bond all miners together.

Brave Men of Gleision

The Tawe meandering, swiftly with ease,
evergreens rustling, in a light breeze.
Birds sweetly singing way up on high,
Cilybebyll, not knowing, brave men would die.

Seven men drifted, ninety metres below,
routine blasting, the way it should go.
At nine twenty-one, all went tragically wrong,
a torrent of water, released on the throng.

The volume of water, a dam being breached.
three of the seven, the entrance they reached.
Rescuers hampered, divers sent down,
the following day, all four souls found drowned.

A tragedy, the worst for decades,
united a village, those hard times ahead.
Hearts broken in pieces, on that fateful day,
love and support, the Welsh mining way.

The Tawe still flowing not far from the mine,
Cilybebyll, will mourn, with love so divine.
Lest we forget, brave souls that were lost,
mining for coal, comes at a great cost.

In memory of the four brave miners who lost their lives on the morning of the 15th September 2011, at the Gleision drift mine, Cilybebyll. Rest in Peace.

Cathedral

My soul is entombed, a cathedral below,
I still walk the face, having nowhere to go.
On the day of the fall, I was sitting alone,
my 'butties' were there, our shift was all done.

The props started creaking, but I paid no heed,
then a noise like thunder, I shielded my head.
The roof started caving, I ran back down the face,
black dust and rock, all giving chase.

I then had to stop, I had no place to go,
curled up in a ball, deadly silence would follow.
Kept my eyes closed, praying, 'God let me live',
life flashing before me, 'twas then 'I Believed'.

I opened my eyes, pitch black, no light,
the face then lit up, a fireball in full flight.
Consuming my air, lungs bursting and burning,
I just passed out, 'twas the end of my journey.

No return to the surface, my tomb now below,
my soul always lingering, with no place to go.
I pray each day, in the Cathedral below,
not for myself, but for other lost souls.

The Runaway

Under the screens the deed would be done,
a truck runs away, at least 20 ton.
The brake released by someone unknown,
away down the track at speed it would go.

Crowds gathered quick, the bounty in site,
coal for the families, to help with their plight.
'Don't let it pass' a voice loud would shout,
change the points quick, time's running out.

Points changed as ordered, the truck losing speed,
straight up the gradient, buffers ahead.
Then with a crash, the truck gets derailed,
toppling over, the bounty revealed.

Barrows and buckets, out of the blue,
coal duly picked, like ants, it's their due.
Lookouts kept watch for coppers a plenty,
by the time they arrived, the truck it was empty.

Times were hard in the valleys those days,
coal that was scavenged, kept cold nights at bay.
Back to the future, with my cover now blown,
Yes, you have guessed, I was that unknown.

Kellingley – Last of an Era

The death knell of mining, a sad fateful day,
no more deep mines, it's the Conservative way.
The industry killed off, by Thatcherites,
four fifty men, now what of their plight.

Hancock the man who put paid to a culture,
he gave no reprieve, like Maggie the vulture.
Cheap coal from abroad, is the way that they think,
no thought for the families, now on the brink.

They argued and fought, their pit to save,
the 'Beeston Seam' had years left to give.
UK Coal, would have given three more,
Hancock ignored this and just slammed the door.

The vulture has said, twelve weeks you're worth,
for coughing and digging in the bowels of the earth.
Four hundred and fifty redundant, with the stroke of a pen,
grown men crying, what's left for them.

I hope in the future, the past will return,
mines will re-open, our coal again burn.
Coal from abroad, is cheap at this time,
but will it last, there's no reason or rhyme.

Miners of KELLINGLEY, I feel your sorrow and pain,
I hope like a phoenix you will rise up again.
To the men and their families, I wish you good cheer,
I hope that the future brings solace not fear.

Kellingley Colliery closed on 18th December 2015, marking the end of deep mining in the United Kingdom.

Now But a Memory – Caerau

The village once vibrant, work did abound,
three times a day, the hooter would sound.
Looked up the lanes, like black ants they marched,
covered in dust, dry mouthed and parched.

Halcyon days in the valley of coal,
brave men gave their lives, bodies and souls.
Childhood memories, so vivid and bright,
the miners on top, a wonderful sight.

They worked hard, drank, lived life to the full,
hard men they were, talked with no bull.
My dad, small, sinewy and strong,
in his mind's eye, miners could do no wrong.

Let's not forget, the mothers and wives,
nurtured the kids, all through their lives.
A breed that would stand up and fight,
put up with so much when money was tight.

I returned to my village a short while ago,
gone are the pits, there's nothing to show.
Boarded up shops, all on the dole,
a community ravaged, now with no soul.

Senghenydd Remembered

The Baron of Merthyr is who he became,
he is the one who must shoulder the blame.
Told about safety, but paid no heed,
driven by profit and pure greed.

Owned Universal, at the time of the deed,
no safety in place, miners would plead.
Mine owners, all the same at this time,
miners like rats, when they went down the mine.

An October morning, brought it all to a head,
439 brave men would all end up dead.
The methane ignited, caused by a spark,
the fire would kill them alone in the dark.

We cannot imagine, the hell they went through,
coughing and choking, if only we knew.
The Baron of Merthyr and cohorts got fined,
a pittance for murder, their pockets all lined.

The life of a miner, meant little at that time,
it was all about profit and that was a crime.
Thirty-four pound was the fine he was given,
no feeling, remorse for the dead or the living.

R.I.P. miners of Senghenydd

The Sinking Pit

Flames raged and licked, that day in the 'Glyn'
brave men paid the price, for coal held within.
Fire damp the cause, that fateful day,
no bodies recovered, they rest where they lay.

The rescue was futile, a funeral pyre,
'The Glyn' would be flooded, to put out the fire.
A few days later, more grief would be spread,
a fireman's error, would leave many more dead.

The sinking pit that was the lure,
employed many men, future secure.
That February morning, the devil and dread,
just like 'The Glyn' men buried and dead.

The men all went down, to that dark silent place,
not knowing their fate, with a smile on their face.
The fears of families in the back of their minds,
not ever thinking, this could be the last time.

They say that the gas had come from 'Glyn Pit'
it travelled a mile, silent and swift.
The explosion occurred down at Cook's slope,
vast, loud and deadly, left little hope.

A mother would cry, my three have been taken.
her grief would be felt, by a whole mining nation.
One sixty-seven was the number they lost.
a fireman's error and they all paid the cost.

In memory of all the lost souls at Llanerch, who died on 6th February 1890. Rest in Peace.

Ynyscedwyn-Hendreladus Colliery Disaster, 24.08.1892

Loved ones they kissed, that tragic night,
their fate in God's hands, alone in their plight.
Three souls survived, six lost their lives,
heartbreak and tears for lovers and wives.

Brothers in arms at the mouth of the drift,
brave to a man, at the start of their shift.
Steel ropes secure, their life line in place,
into the 'bridies' and down to the face.

A clutch disengaged, an engine unmanned,
the 'bridies' careered, a future unplanned.
Down the slope speeding, out of control,
violently crashing, taking bodies and souls.

Pit bottom they lay, bodies all broken,
battered by 'bridies' the Devil awoken.
Steel against flesh, not a fair fight,
a miracle to survive, without a doubt.

Mines now long gone, an industry dead,
disasters like this, were always a dread.
Let's never forget those brave men of coal,
who gave up their lives, bodies and souls.

Today on the site, laughter abounds,
children are taught about life underground.
The legacy left by the six souls below,
lives on forever, these children will know.

The Feeder and Suckers

Don't go near the feeder my dad would say,
the suckers will get you, there'll be hell to pay.
The three of us listened, just used to gaze,
other kids stripped and swam in the haze,

Dragonflies hovered, kids splashed and played,
not realising the danger, completely unfazed.
Being so young, not a care in the world,
paying no heed to what my dad had said.

The feeder was calm, then the suckers did roar.
A life would be taken, the village would mourn.
I remember it well, that afternoon,
splashing and laughter, when the fans they did turn.

One minute he's there, the next he is gone.
the suckers had got him, the deed had been done.
Took ages to silence, by miners on top.
on the bank crying, tears never stopped.

The doctor was summoned, Jill was her name.
dived in to the feeder, a body re-claimed.
All effort was made to save his young life.
too late however, amongst all the strife.

A child's life was lost in a heartbeat.
vows later made, there'd be no repeat.
The words of my dad, still haunt me today.
'The suckers will get you, there'll be hell to pay.'

Put the Coal Away

Once a month dropped in the lane,
the N.C.B.'s loss, our families gain.
A ton for almost every house in the street,
to keep us warm and give us heat.

Mam would say 'Put the coal away',
my brothers and I would duly obey.
We opened the cwtch and just threw it in,
when we finished, there's no longer a din.

The hooter then blew, we ran to the pit,
met my dad, thought we were it.
We've chucked the coal in, dad we said,
he turned, smiled and just nodded his head,

Down the lane, smiling so proud,
Dad coughing, that horrible sound.
He opened the cwtch, and said 'Oh my God',
you've put it in wrong, a terrible job.

Within the hour, he'd emptied the lot,
then put it back, thought he'd lost the plot.
That's the way to put coal away,
now fill my bath, it's been a hard day.

Traits of a Miner

Sinews and muscles, a miner's mainstay,
pushed to the limit, day after day.
Sweat cascading, a contorted face,
putrefied air, a dark dingy place.

The mine it became, a mistress for life,
loved her to death, through trouble and strife.
Resilience and courage they had on their side,
blue scars they wore, with honour and pride.

Played and drunk hard, a pure fallacy,
kind men and fair, the reality.
My dad was one of this special breed,
always around in the hour of need.

'Twas in the blood to go underground,
father to son, a career handed down.
A mining legacy, now long redundant,
a government's doing, oh so repugnant.

The Knockers… BWCA

Malevolent spirits, roaming deep underground,
mythical creatures, their loud 'knocking' sounds.
Two feet tall, with mischief their calling,
miner's garb worn 'knocking' their warning.

Pit props would creak, a miner's nightmare,
methane and rocks, a swift deadly pair.
'Knockers' would warn, danger below,
brave miners would run, evading the foe.

They wander the face, in the bowels below,
playing their tricks, causing miners much woe.
Food they would steal, as if by right,
not seen or heard, like thieves in the night.

Many believed the 'Knockers' existed,
others more sceptic, their warnings resisted.
Death their reward, for mining below,
buried alive, now with no place to go.

Who are these 'Knockers' lurking below,
dead miners they say, or maybe their souls.
Their wicked deeds, folklore history,
'Knockers' still shrouded, in deep mystery.

Disaster in the Trefor

The nightshift did warn, of the dangers that lurked,
a cocktail of death, in the bowels of the earth.
Methane and rock, would kill nine brave men,
a fireball would take them, in heaven they dwell

A few weeks before a 'jump' fault was found,
to re-join the seam, steel rings would abound.
Most men withdrew, safely that day,
pockets of methane, like assassins they lay.

Another rock fall, a steel ring and spark,
ignited the methane, a fireball embarked.
In an instant two taken, no time to retreat,
seven butties burnt, later deceased.

The explosion was felt, miles underground,
in the 'Bertie' I heard, that deathly knell sound.
I remember the 'whoosh' just like a tube train,
that day's memories will always remain.

Five men survived, the blazing inferno,
each to a man, an underground hero.
Life is so random, when digging for coal,
let's never forget, all those lost souls.

Dedicated to the nine brave miners who lost their lives in the Lewis Merthyr pit disaster, 22nd November 1956.

Weekly Card School... Albert Street, Caerau

Always on a Sunday, around about nine,
a card school would open, just down from the mine.
The deck would come out, miners they'd gather,
now and again including my father.

Look outs required to watch for the coppers,
my brothers and I would take up the offer.
Strategically placed on banks upon high,
a whistle the warning, should coppers pass by.

The cap would come out, coin was thrown in,
the gambling began, kith versus kin.
I spotted some coppers, running fast, two abreast,
I whistled the warning, there'd be no arrests.

Quickly the coin and the cap they were gathered,
gamblers bolting, like horses untethered.
After the warning we'd hide in the grass,
covered by ferns as the coppers ran past.

Later that day we'd get our reward.
a tanner each, not bad we all thought,
You saved the day, they'd tell us all three.
the whistle you blew, got us all off scot free.

Same time next week if you fancy it boys,
if you do a good job, it may be two bob.
Today's generation, don't know what they've missed,
growing up in Caerau, behaving like this.

Coppers and miners, like hounds and hares,
watching in the grass... as Max sings... I WAS THERE.

Lord Save Me Now… Arglwydd achub fi nawr

My name is Moses, in the 'Lan' drift I worked,
fire damp and falls, in headings they lurked.
A 'door boy' I was, supplying air for brave men,
our fate in God's hand, time and again.

The 'brass vein' I worked, from morning 'til night,
whilst colliers hewed coal, in naked flame light.
I often got bored, would play 'hide and seek',
then back to the 'door', with a clip for my cheek.

I laughed with the colliers, I respected them all,
without a care in the world, they treated me well.
Taught me about life, deep underground,
brave men and true, their knowledge profound.

The explosion so deafening, came out of the blue,
brave colliers running, a giant fireball ensued.
Knocked off my feet, flame got to me first,
gasping for air, with my lungs fit to burst.

Brave rescuers came, without safety or care,
deadly, noxious vapour permeated the air.
Brave men lay burning, some taken by gas,
bodies retrieved, to the surface en-masse.

'Lord save us now' I heard brave men scream,
then death did befall me, as if all a dream.
At twelve years of age, my maker I met,
my short life in the drift, you must never forget.

I read an article regarding the mining disaster at Lan Colliery, Gwaelod-y-Garth, Pentyrch, on the morning of 6th December 1875, when twelve men and boys were killed. One of the boys was Moses Llewellyn, a 'door boy'. I understand there is going to be a film made about it in the future.

Tip Number Seven

Tip number seven erupted that day,
a man-made disaster, a village would pay.
Coal waste and sludge, a dying legacy,
its wrath did rain down, Aberfan's destiny.

Children so happy, 'twas half term that day,
a week with no school, hip, hip hooray.
Playing in orchards and catching tadpoles,
not a care in the world, but sadly lost souls.

A grey mist lay thick in the valley that morn,
just a few hours later, many families forlorn.
Why did it happen, the truth never known,
heartbroken parents, in sorrow still drown.

Lest not forget the adults that perished,
protecting the young, lives that they cherished.
Their calling in life, to nurture and mould,
their calling in death, bravery to behold.

Fifty years later, their memory lives on,
a generation lost, on that autumnal morn.
Innocent children, before their time taken,
now in God's care never forsaken.

> *It was this tip that was responsible for the Aberfan disaster. It was man-made and the National Coal Board and politicians reneged on their responsibilities. Shame on them all.*

Tragedy in the 'Silkstone Seam'

That day in the 'Silkstone' never forgotten,
brave miners dying, below on pit bottom.
Shot firing as normal, going as planned,
then, on the tenth 'Satan's' command.

Methane ignited, fire, dense smoke and fumes,
with no place to hide, all air consumed.
The fireball and blast, no forgivers of life,
on a path all consuming, devastation now rife.

Five heroes were taken, in the darkness below,
one thing on their mind, the search for 'Black Gold'.
Loved ones left grieving, lives never the same, no fault was
accorded, blame methane and flame.

This brave breed of men, now confined to history,
'What drives them on?' it's a complete mystery.
Only those who have hewed, down that dark hole below,
can give us the answer, with their 'blue scars' on show.

The Allerton Bywater Mining Disaster – In memory of George Paley, Arthur Richards, Albany Taylor, William Townsend and John Allan, who lost their lives on 10th March 1930. May they all Rest in Peace.

Keeper of the Collieries

A once vibrant valley, coal its heartbeat,
sadly forsaken, barren and bleak.
Collieries sunken, never to return,
a community heartbroken, still grieve and yearn.

What of the future, how do you replace?
A legacy of mining, vast empty space.
With a woodland of spirit, abundant blooms,
overseen by the 'Keeper' erasing the gloom.

Carved out of oak, the 'Keeper' stands proud,
evoking memories, brave men underground.
"The Keeper of Collieries" reminds of the past,
an industry lost, once a proud beating heart.

The Llynfi valley, blossoming once again,
on a site, where black gold, surrendered his reign.
Cherish the 'Keeper', many departed souls,
those brave men who mined, in the bowels below.

The Keeper of the Collieries is a nine-foot tall oak sculpture depicting a miner. He has been placed in the middle of a new community woodland area overlooking the Llynfi Valley, a once vibrant mining valley, sadly no more. His face appears on the cover of this book.

I Wander the Face

A 'drawer' I was in that black hole below,
chained by the waist, my life full of woe.
I pulled carts of coal, twelve hours a day,
a slave to my master, I duly obeyed.

At eight years of age, my future so bleak,
bare chested, shackled, solace I'd seek.
My life of no value, I lay down to die,
coal dust engulfing, as I closed my eyes.

Worked to the bone, exhausted, no air,
my life ebbed away, as I muttered a prayer.
Now I wander the face, no hope of release,
my spirit entombed, an unbridled peace.

Superstition is rife, in the bowels below,
spectres appear, then in seconds they go.
I cannot be seen, but I'm there to protect,
those brave souls below, their saviour elect.

When danger is lurking, I give them a sign,
the clank of my chains, their ghostly lifeline.
Miners believe, dead souls wander the face,
I am just one, I am there by God's grace.

> *In the Victorian era, young children were made to work in the mines. Conditions were horrendous and many died underground. I penned this poem and incorporated it with the superstition of 'spectres' who would warn miners of impending danger... Are you a believer?*

Always in our Prayers

Devastation dawned early, that autumnal morn,
no warning given, a village would mourn.
A destructive torrent, slag, slurry and coal.
'Pant Glas' engulfed, many angelic souls.

The whole world looked on, in total despair,
a Welsh mining village, united in prayer.
Brave miners rallied, rescuing with haste,
children and staff, from a black deadly waste.

With bare hands they dug, searching for life,
whilst villagers wept, through all the strife.
Lost souls that day, fought for life saving air,
a black rancid slurry, laid classrooms bare.

As day turned to night, with hope fading fast,
rescuers exhausted, a dark shadow cast.
On that fateful day, the Lord took one forty four,
if only he'd given them, a few hours more.

That deadly morning, all laughing, nothing amiss,
without any worries, their young lives were bliss.
Not a care in the world, half term that day,
a few minutes later, innocence stolen away.

'Aberfan' has survived, scars have now healed,
'The Lord' did give life, grief never concealed.
They say that time heals, you may not agree,
fifty one years on, we're all still shedding tears.

To commemorate the 51st anniversary of the Aberfan disaster, Friday 21st October 1966. Never to be forgotten.

Was it Their Destiny?

A breed set apart, with courage to spare,
resilience, fortitude, sometimes despair.
Blue scars they wore, like badges of honour,
in the valleys of coal, no bond was stronger.

Mining black gold, life's only purpose,
a steel cage their lifeline, back to the surface.
To enter the abyss, brave men indeed,
following fore-fathers, as if by decree.

On bellies they'd crawl, day after day,
breathing black dust, lungs decaying away.
In this kingdom below, coal was the King,
when stirred in anger, death he would bring.

A risk they all knew, ignored out of hand,
this mindset however, we can't understand.
Unless we have mined, in the bowels below,
the answer lies there, in body and soul.

Valleys they thrived, when coal was aplenty,
pits their heartbeat, now bleak and empty.
Miners reminisce, hearts broken with sorrow,
if truth be told, they would go back tomorrow.

We must never forget, their lasting legacy,
maybe from birth, it was true destiny.
When a miner you meet, just ask him why,
he will answer sincerely, with a tear in his eye.

> *I wrote this poem with my old mate Steve Johnson in mind, he being an ex-miner. To say these brave men are a rare breed is an understatement, and I have often wondered what drew them to the black abyss below. I am sure there are many reasons.*

Father and Son in Spirit

As the son of a miner, my life was mapped out,
to follow his footsteps, it was never in doubt.
At the age of fourteen, I was a boy amongst men,
on bellies we'd crawl, 'Black Gold' beckoning.

For ten years I toiled, in the bowels below,
stripped to the waist, daily baring my soul.
At my father's right hand, protected each day,
in the dust and the gloom, daily we'd pray.

When having our snap, many stories were told,
of folklore and spirits, my blood running cold.
My workmates would tease, it's an underground culture,
my father reminding me, son this is your future.

I was sixteen years old, when my father passed,
crushed in a fall, it all happened so fast.
Without any warning, the roof just collapsed,
his broken body I cradled, as his last breath elapsed.

Brought to the surface, hooter still blasting,
my ashen faced mother, her grief everlasting.
My heart was broken, on that early morn,
my hero, my soulmate, I was left so forlorn.

After the wake, it was back to the face,
conditions so grim, I survived by God's grace.
Methane all run, was the last thing I heard,
a fireball consumed me, I never uttered a word.

Life flashing before me, my father appeared,
like a beacon of light, saying join me, don't fear.
With both hands outstretched, he beckoned to me,
don't be afraid, the Lord's set your soul free.

Now I wander the face, protecting brave men,
overseeing their teasing, as it was way back then.
When danger is lurking, I give them a sign,
am I spirit or folklore, or a power divine.

> *I have read so many mining stories of spirits and ghosts protecting brave miners and giving them warning of imminent disaster, way down below in the bowels of the earth. I thought I'd pen a poem for them.*

Pit of Nations – Gedling Colliery

From sunny climes they came, to seek a better life,
homelands now forsaken, abandoning all strife.
In Nottingham they settled, welcomed with open arms,
life would not be easy, many storms before the calm.

What would the future hold, local tempers running high,
black foreigners upon our shores, 'racism' was the cry.
Work was at a premium, young mouths they had to feed,
'Black Gold' became their calling, desperate in times of need.

They worked the 'Pit of Nations' courageous to a man,
descending to the bowels, decades it did span.
A band of brothers joined, colour it meant nothing,
comrades they became, no thought of hate or loathing.

That deep black hole they toiled, in racial harmony,
on knees and bellies crawling, hewing tirelessly.
Mining was hard and dangerous, but there were no regrets,
however once back on top, racist taunts and threats.

Over time they were accepted, black became the norm,
these Industrial pioneers, had weathered all the storms.
A journey from a land of sun, to the darkest place on earth,
where humour did abound, 'faces' filled with mirth.

Our coalfields now discarded, confined to history,
no mines producing 'Black Gold' a political decree.
Black miners reminisce, trailblazers to a man,
it was in the 'Pit of Nations, where it all began.

I wrote this poem after reading an article regarding black immigrants who landed on our shores in the 1950's from the Caribbean. Many settled with their families in Nottingham and, although they were welcomed by the majority of the community, they were subjected to hate campaigns by a minority. Work was at a premium so many became miners at Gedling Colliery a.k.a. 'The Pit of Nations'. Underground they were treated as equals, they were a band of brothers. Bearing in mind none of these men had ever seen a coal mine, they were regarded as pioneers in the industry.

Knockshinnock Castle Colliery Mining Disaster

Many years it lay dormant, a vast glacial lake,
nature's creation, thirteen lives it would take.
A torrent unleashed, leaving death in its wake,
a river of peat, causing pain and heartache.

Trapped deep in the bowels, a hell hole below,
to the surface no egress, they had nowhere to go.
An entombed 'band of brothers' all praying to God,
their hope to be rescued, against all the odds.

Awash with sludge, escape routes all sealed,
a true 'shift of heroes', who never did yield.
With fresh air polluted, fighting to survive,
divine intervention, would keep them alive.

Time at a premium, a rescue was mounted,
brave men on top, stood up and were counted.
Their mission to save, those imprisoned below,
to unite their families, a precious gift to bestow.

Gas filled old workings, death traps in waiting,
rescuers hindered, loved ones hoping and praying.
After digging for days, contact was made,
a rescue success, so much courage displayed.

Once on the surface, breathing fresh air so sweet,
prayers had been answered, loved ones they'd greet.
Celebrations subdued, 'thirteen brothers' entombed,
many weeks it would take, before finally exhumed.

In the bowels below where day becomes night,
rescuers and survivors, hailed heroes by right.
In the belly of the beast, not knowing their fate,
this 'mistress for life' only they could relate.

> *At 7.30pm, Thursday 7th September 1950, a large volume of liquefied peat flooded the workings of Knockshinnoch Castle Colliery, Ayrshire, trapping 129 miners and killing 13. The ensuing rescue lasted nearly three days and oxygen re-breathers were used to get the survivors out. Months later the bodies of the deceased were recovered. A truly remarkable series of events, showing how dangerous mining is and the courage that miners need when digging for 'Black Gold'.*

The Eden Colliery – Place of Pleasure

Coal mining to me, was my genetic calling,
'twas the pit, from the day I was born.
My life was mapped out, before even crawling,
I delved in the bowels, night and dawn.

Sixteen I was when I started,
from a boy to a man in no time.
Working with men, so stout hearted,
some lives cut short by dust, in their prime.

On bellies we'd crawl, in the darkness,
camaraderie was second to none.
If life was lost, eyes filled with sadness,
a band of brave brothers, rallying as one.

Burrowing like moles, dust kissing our lips,
night and day, in the belly of the beast.
A world underground, like a total eclipse,
bread, cheese and onion, our daily feast.

'Eden' they say, a wondrous 'place of pleasure',
miners bending their backs, the actual reality.
In search of black gold, their ultimate treasure,
blue scars reminders, of surviving mortality.

For thirty-four years, I was a subject of 'Eden',
she was my queen, I served her well.
Some friends I lost, no rhyme or reason,
that 'place of pleasure' on occasions was hell.

In July 1980 'Eden's' seams finally played out,
my queen abdicated, dust kissed me no more.
So much sadness, of that there's no doubt,
I would do it all again, like my father before

Today is a day of remembrance,
these memories are with me for life.
My industry destroyed, political arrogance,
causing so much heartache and strife.

To commemorate the 38th anniversary of the closure of Eden Colliery, Leadgate (18th July 1980). I've written it through the eyes of an 88-year-old miner who spent 34 years underground there.

A Pony's Tale

A free spirit I was, back there in the day,
I worked the face, but often I'd stray.
My halter I'd slip, when stabled below,
How I survived, our dear Lord only knows.

I wandered, no cares, alone in the dark,
then a thunderous noise, ignited by spark.
A resounding roar, I had nowhere to go,
alone down the face, in the bowels below.

Stablemates banished, death in an instant,
thoughts in my head, only future existence.
To the stables I wandered, breathing in dust,
my lungs fit to burst, fresh air I did lust.

For days I would wander, with no sign of life,
no hope of a rescue, all alone in the plight.
To survive I foraged, on comrades so brave,
my life they did save, beyond the grave.

For two months I survived, how God only knows,
twenty-one ponies, four brave miners' souls.
Slipping my halter, saved from certain death,
surviving on comrades, who gave me the strength.

> *In 1808 an explosion took place in Harraton Colliery, Durham; twenty-one ponies and four colliers were killed, and just one pony survived. He was found after being entombed below for two months.*

Hope... Gleision

A village united, hearts filled with grief,
beloved menfolk, alone in the deep.
They 'drifted' that morning, without a care,
Coal was their goal, hours later they despaired.

Four brave miners, friends to the last,
never forgotten, bravery unsurpassed.
Twenty-four hours, hoping and praying,
Cilybebell united, as loved ones lay dying.

The moment of truth, rescuers' nightmare,
survival impossible, Cilybebell's grief shared.
Until that minute, hope sprung eternal,
sadness a shroud, the 'drift' now infernal.

The mountain it claimed, four heroes that day,
black gold their master, with life they did pay.
Gleision a shrine, to brave men and true,
never forgotten, our loves last 'adieu'.